HOLZ books | 2018

SOLAR SYSTEM

What is the solar system? It is our Sun and everything that travels around it. The Sun is in the center of the solar system. Our solar system is always in motion. Eight known planets, along with comets, asteroids, and other space objects orbit the Sun. The Sun is the biggest object in our solar system. It contains more than 99% of the solar system's mass. Nearest the Sun are four fairly small, rocky planets - Mercury, Venus, Earth and Mars.

On the far side of the asteroid belt are the four gas giants – Jupiter, Saturn, Uranus and Neptune. These planets are much bigger than Earth, but very lightweight for their size. They are mostly made of hydrogen and helium. Until recently, the furthest known planet was an icy world called Pluto. However, Pluto is dwarfed by Earthes Moon and many astronomers think it is too small to be called a true planet.

Jupiter

Mercury Venus Earth Mars

SUN

The Sun (or Sol), is the star at the centre of our solar system and is responsible for the Earth's climate and weather. The Sun is an almost perfect sphere with a difference of just 10km in diameter between the poles and the equator. The average radius of the Sun is 695,508 km (109.2 x that of the Earth) of which 20–25% is the core. The sun was born about 4.6 billion years ago. Many scientists think the sun and the rest of the solar system formed from a giant, rotating cloud of gas and dust known as the solar nebula. As the nebula collapsed because of its gravity, it spun faster and flattened into a disk. Most of the material was pulled toward the center to form the sun.

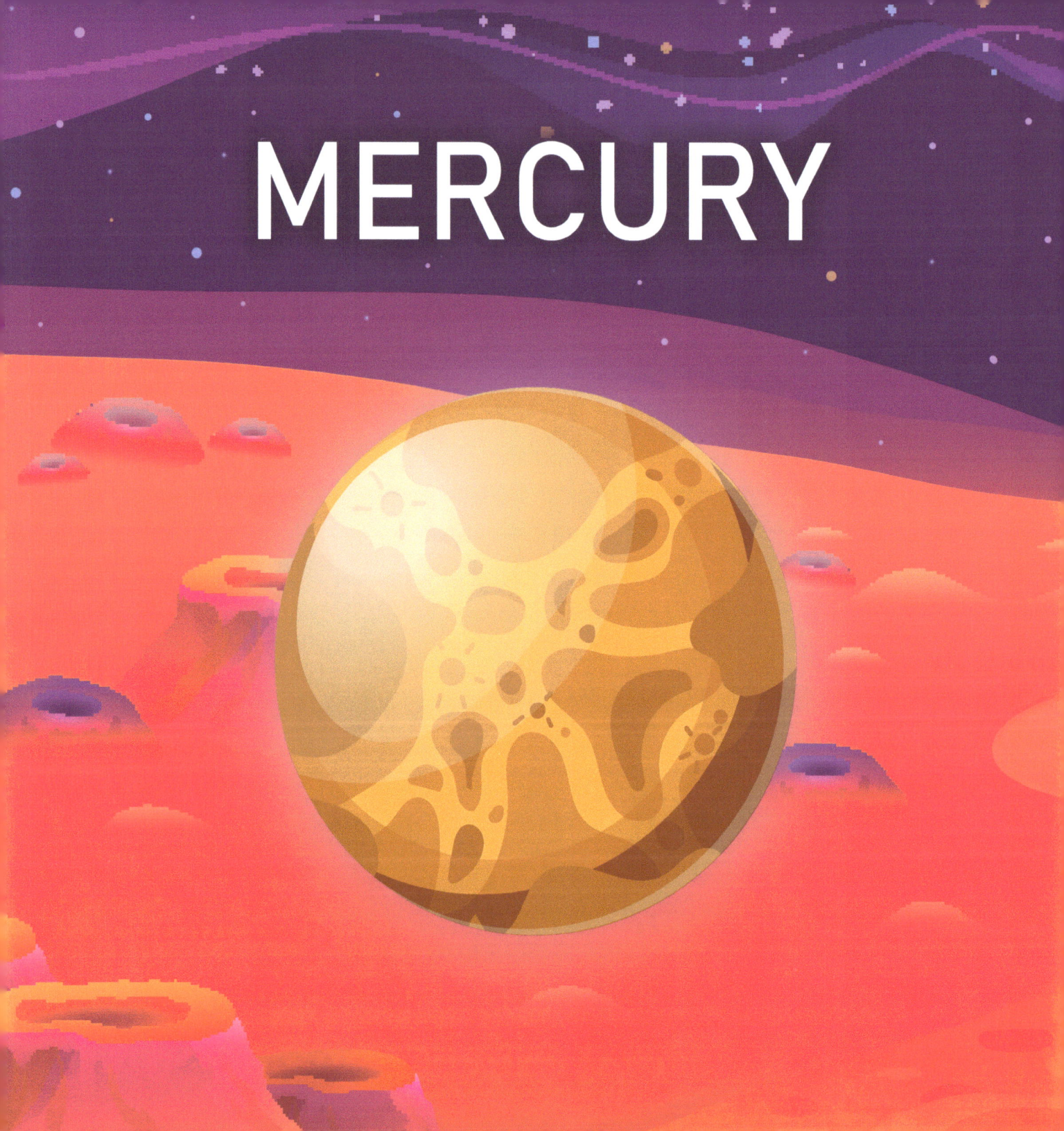

Mercury is the first planet from the Sun. Mercury speeds around the Sun once every 88 days. However, it spins on its axis very slowly — once every 58.6 days. This is exactly two thirds of its orbital period. Mercury is a small, rocky world. The planet Mercury is just a ball bearing of iron, with a thin layer of rock. About 80% of the diameter of the whole planet is taken up by its core of metallic iron. For some unknown reason, Mercury is linked to the Earth. Every time Mercury reaches its closest distance to us (every 117 days) it always shows the same face to us. Mercury was born in giant collisions about four and a half billion years ago, as huge rocks collided together.

VENUS

Venus is the hottest planet in our solar system. This hostile world is covered in thousands of volcanoes and is encased in a dense layer of toxic clouds, swept along by constant hurricane-force winds.

On Venus, the clouds cover 100% of the planet all of the time. The clouds begin at 50 km above the surface, and then continue for another 25 km above that. The clouds are made of droplets of sulphuric acid, which are about 50 times smaller than the thickness of a human hair. Venus takes 225 earth-days to make a complete loop around the Sun. In fact, it rotates very slowly, and a Venus Day lasts 243 Earth Days. So on Venus a day is longer than a year.

Our home planet Earth is a rocky, terrestrial planet. It has a solid and active surface with mountains, valleys, canyons, plains and so much more. Earth is special because it is an ocean planet. Water covers 70% of Earth's surface. With a radius of 3,959 miles, Earth is the fifth largest planet in our solar system, and it's the only one known for sure to have liquid water on its surface. Earth orbits the sun once every 365.25 days. Since our calendar years have only 365 days, we add an extra leap day every four years to account for the difference. These bodies of water contain 97 percent of Earth's volcanoes and the mid-ocean ridge, a massive mountain range more than 40,000 miles long.

Mars is a cold desert world. It is half the size of Earth. Mars is sometimes called the Red Planet. It's red because of rusty iron in the ground. Like Earth, Mars has seasons, polar ice caps, volcanoes, canyons, and weather. Mars is the fourth planet from the Sun. The composition of Mars' atmosphere is extremely similar to Venus. The main component in both atmospheres is carbon dioxide (95% for Mars, 97% for Venus), yet a runaway greenhouse effect has taken hold of Venus, producing temperatures in excess of 480° C, while temperatures on Mars never exceed 20° C.

JUPITER

The largest planet in our solar system, Jupiter is a giant ball of gas 300 times more massive than earth. It's five times farther from the Sun than we are. A year on Jupiter is twelve times longer than ours, but a Jovian day is only ten hours long. Jupiter's appearance is a tapestry of beautiful colors and atmospheric features. Water exists deep below and can sometimes be seen through clear spots in the clouds. The planet's "stripes" are dark belts and light zones created by strong winds in Jupiter's upper atmosphere. The Great Red Spot, a giant spinning storm, has been observed for more than 300 years. It is very much like a hurricane on Earth except that it is large enough that three Earths would fit within its boundaries.

# SATURN

Saturn is the sixth planet from the Sun and the most distant that can be seen with the naked eye. Saturn is the second largest planet and is best known for its fabulous ring system that was first observed in 1610 by the astronomer Galileo Galilei. Like Jupiter, Saturn is a gas giant and is composed of similar gasses including hydrogen, helium and methane. Saturn orbits the Sun once every 29.4 Earth years. Its slow movement against the backdrop of stars earned it the nickname of "Lubadsagush" from the ancient Assyrians. The name means "oldest of the old".
The Saturnian rings are made mostly of chunks of ice and small amounts of carbonaceous dust. The rings stretch out more than 120,700 km.

URANUS

Uranus is the seventh planet from the Sun and the third largest (by diameter). Like the other gas planets, Uranus has bands of clouds that blow around rapidly. Uranus is blue-green in color, the result of methane in its mostly hydrogen-helium atmosphere. The planet is often dubbed an ice giant, since 80 percent or more of its mass is made up of a fluid mix of water, methane, and ammonia ices. Uranus has the coldest atmosphere of any of the planets in the solar system, even though it is not the most distant. Despite the fact that its equator faces away from the sun, the temperature distribution on Uranus is much like other planets, with a warmer equator and cooler poles.

Neptune is the eighth planet from the Sun making it the most distant in the solar system. This gas giant planet. Neptune spins on its axis very rapidly. Its equatorial clouds take 18 hours to make one rotation. This is because Neptune is not solid body. Neptune has 14 moons. The most interesting moon is Triton, a frozen world that is spewing nitrogen ice and dust particles out from below its surface. It was likely captured by the gravitational pull of Neptune. It is probably the coldest world in the solar system. If Neptune were hollow, it could contain nearly 60 Earths. Neptune orbits the Sun every 165 years.

PLUTO

Pluto is a dwarf planet that lies in the Kuiper Belt. It's an area full of icy bodies and other dwarf planets at the edge of our solar system. Pluto is the biggest object in this region. Since Neptune's gravity influences its neighboring planet Pluto, and Pluto shares its orbit with frozen gases and objects in the Kuiper belt, that meant Pluto was out of planet status and was reclassified from a planet to a dwarf planet. Pluto is one third water. This is in the form of water ice which is more than 3 times as much water as in all the Earth's oceans, the remaining two thirds are rock. Pluto's surface is covered with ices, and has several mountain ranges, light and dark regions, and a scattering of craters.

COMET

Comets are actually invisible until they begin to get close to the Sun. As they begin to heat up, an amazing transformation takes place. The dust and gases frozen within the comet begin to expand and burst forth at explosive velocities. The solid part of the comet is called the nucleus, while the envelope of dust and gas around it is known as the coma. Solar winds cause the material in the coma to trail behind the comet for a much as a million miles. As the Sun illuminates this material, it begins to glow brightly. This forms the famous tail of the comet. Comets and their tails can usually be seen from Earth and can be quite bright if conditions are right. Some comets may have as many as three separate tails.

www.holzbooks.com